Contents

What is a force?

Forces make things happen. Think of a banana sitting in a fruit bowl. Suddenly a force lifts it into the air – that's your hand at work. Another force pulls its peel away – your hand again.

People have always been amazed by the way winds blow, lightning flashes, and earthquakes and tsunamis flatten houses. Why do apples fall downwards, not sideways? Why is iron attracted to magnets? What causes all these mysterious forces?

Looking for the answer

People used to believe that forces were caused by gods and spirits. An ancient Greek thinker, **Aristotle**, came up with another explanation. He thought forces were caused by the flowing of four basic ingredients of the Universe: earth, air, fire and water. So a flame reached upwards because it wanted to return to the Sun; while apples, which came from the Earth, fell to the ground in autumn.

Aristotle's theory was the accepted view for 2000 years, and was of course completely wrong. Our understanding of forces and how to use them has been slow in coming.

Right: Aristotle lived in the 4th century BCE and was one of the first scientists. He studied forces, weather, animals, human senses and many other things.

Magic pendulum

Place a number of magnets on a table, under a piece of paper. Tie the end of a piece of string around another magnet. Let the magnet swing above the paper like a pendulum. Because of the forces from all the separate magnets, it will sway and wobble as if controlled by magic.

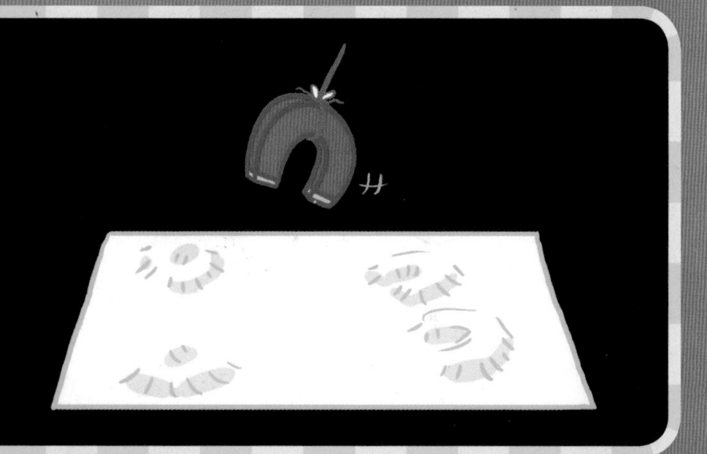

Super Science

Forces and Movement

Richard Robinson

QED Publishing

First published in the UK in 2007 by
QED Publishing
A Quarto Group company
226 City Road
London EC1V 2TT
www.qed-publishing.co.uk

A catalogue record for this book is available from the British Library.

ISBN 978-1-84538-913-0

Written by Richard Robinson
Edited by Anna Claybourne
Designed by Balley Design Ltd
Consultant Terry Jennings

Publisher Steve Evans
Creative Director Zeta Davies
Senior Editor Hannah Ray

Printed and bound in China

Picture credits
Key: T = top, B = bottom, C = centre, L = left, R = right, FC = front cover

Bridgeman Art Library: p22 William of Tyre/Bibliotheque Municipal de Lyon, France.

Corbis: p4 Bettmann; p13 S.Feval/Le Matin/Sygma; p14 Paul A.Souders; p19 Hulton-Deutsch Collection; p25 Bettmann; p27 Chase Jarvis.

NASA: p6 p10 ISS; p14 Langley Research Center.

Science Photo Library: p8 p11 NASA; p17 Keith Kent; p18 Andrew Lambert Photography; p20 Volker Steger; p29 David R. Frazier.

Words in **bold** can be found in the Glossary on page 31.

Pulling from a distance

Magnetism is a distant force. Magnets can pull without touching. A magnet will pull a steel paperclip towards it.

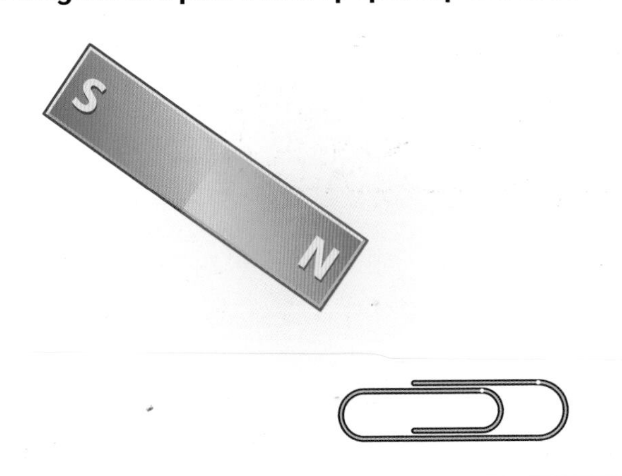

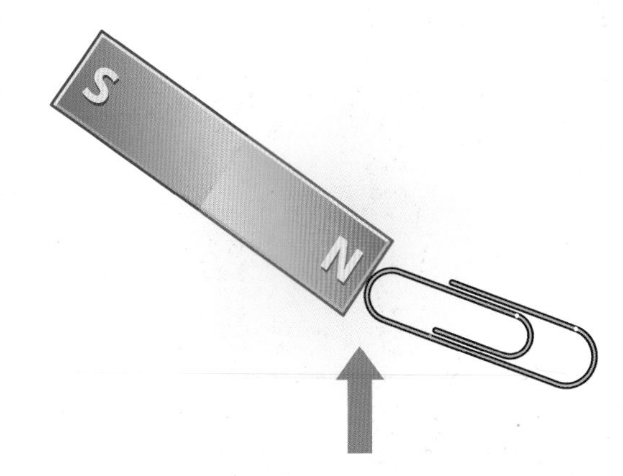

Types of forces

There are two main kinds of force:

Contact forces involve things touching each other, and pushing or pulling against each other. Kicking a ball is a good example.

Distant forces are invisible forces, such as **gravity** or magnetism, that can work on an object without touching it.

As well as exploring the different kinds of forces, this book is about how we harness forces and use them to do things for us.

Do you know of any other strange things ancient peoples used to believe about the Earth? See page 30.

Above: People used to think that gods caused earthquakes, lightning and tides. Zeus was the leader of the ancient Greek gods.

Newton's laws

Our understanding of forces today is based on the work of a great genius called Isaac Newton, who lived from 1643 to 1727. So what exactly is a force? A force is a push or pull that makes an object speed up, slow down, change shape or change direction.

Understanding forces

The genius of Isaac Newton was that he worked out precise laws of forces. So precise that today, 300 years later, we use his laws to launch rockets, satellites and space stations. Newton's laws (see opposite) help us predict **eclipses**, the appearances of comets and the passage of stars in the heavens.

Below: The Voyager I space probe is heading out into space beyond the Solar System to explore the Universe. After being launched into space, it used up its fuel. Now, it keeps going because there is no force to stop it. It obeys Newton's First Law (see opposite).

The three laws of motion

Here are Newton's three laws of motion, which are among his most important works. They describe the way forces make objects behave.

Newton's First Law

An object tends to keep moving in the same direction and at the same speed, unless another force acts on it.

Right: If Max is gliding on the ice, he will carry on until he is pushed or bumps into something. Gradually, the forces of friction (see page 8) cause the ice to drag on his skates – which will slow him down.

Newton's Second Law

The effect of a force depends on the **mass** of the object (how heavy it is). It is easier to push a bike than a car, and once they are going it is easier to stop a bike than a car.

Above: Max is bigger and heavier than Minnie, and it is much harder to make him move. Once he gets going, he is harder to stop.

Newton's Third Law

Every action has an equal and opposite reaction. This explains how a rocket works. A strong force pushing backwards from the jets of a rocket pushes the rocket forward.

Right: When Minnie shoves Max away, she is herself pushed backwards.

Friction

As things rub against each other, they slow down. This rubbing force is called friction. For a sledge to take you fast downhill you would want as little friction as possible, but when you got to the bottom you'd need friction to slow you down, otherwise you might go on for ever.

Useful friction

We use friction all the time. For a car to stop, the brake pads clamp against the wheels. Friction between the pads and the wheels gradually makes the wheels stop moving. The rubber on car tyres produces good friction against the road, helping the wheels to grip. But when it rains, water can get between the tyres and the road, making the car skid. The grooves in a tyre are designed to channel water out of the way, so that the rubbery tyre surface can stay in contact with the road.

Below: Even air has friction as it rubs against objects. When a space shuttle returns from space, it rubs against the air, and friction slows it down from its **re-entry** speed of 28 000 km/h to its landing speed of 220 km/h. Friction also causes the shuttle to heat up and glow red-hot.

If it wasn't for friction, a game of snooker would be impossible. The balls would roll around the table for ever. How do you think a friction-free snooker game would end?
See page 30.

Feel the friction

Friction produces noise and heat. Rub your hands together and you will feel the heat and hear the noise.

Above: Tyres are designed to channel water aside so that they stay in contact with the road.

Stopping friction

Sometimes friction causes problems. In a car, friction is needed on the tyres and brakes, but not in the engine. In an engine, friction slows down the parts, wearing them out and wasting **energy**. Friction can be reduced by coating engine parts in smooth, slippery oil.

Playground slides are made of smooth, shiny metal, not rubber or stone. There is very little friction between metal and clothes, so it's easy to slide down the slide.

Make a marble chariot

Place four or more marbles on a table. Then take a small cardboard box, such as the tray part of an empty matchbox, and place it upside-down over the marbles. The sides of the box should not touch the table. (You may need to make a box if you can't find one the right size.) Push your chariot across the table. It should run for some distance. Rolling reduces friction because as an object rolls, it does not rub against the surface it is moving over. Engineers often use ball-bearings to reduce friction between moving parts.

Gravity – an invisible force

One of the most puzzling forces of all is one that is all around us. Gravity is the force that pulls things down to the ground. Although we know a lot about gravity, we have no idea what it actually is. Scientists are still trying to find out!

Pulling force

In everyday life, most people think of gravity as a downwards pull towards the Earth. If you drop something, it is pulled towards the ground. If you jump up, the Earth pulls you back down.

However, gravity isn't just to do with the Earth. Gravity is everywhere, even in space. All objects have gravity – a pulling force that draws them together. The bigger an object's mass (the more matter it contains), the bigger its pulling force.

Even though you are tiny compared to the Earth, you have your own **gravitational field**. While the Earth is pulling you down, you are very slightly pulling the Earth up. Can you think of something that has enough gravity to have a bigger effect on the Earth? See page 30.

Below: In space stations, there is not enough gravity to hold objects down. Food and drinks float about and have to be caught.

Weaker gravity

Because the Moon is smaller and lighter than the Earth, gravity is weaker there, and astronauts can jump much higher than on Earth.

The force of gravity gets less the farther you are from the object. When astronauts are far above the Earth, they escape from gravity's pull.

Left: On a planet with low gravity, a person could jump much higher.

Above: Can you imagine what life would be like in zero gravity?

Ball curve

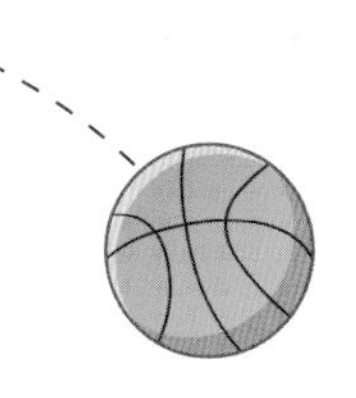

When you throw a ball, you use a pushing force to propel it into the air. The pull of gravity makes it curve back towards the ground.

Pressure

Any object that pushes against another object is applying pressure. There's also pressure all around us in the form of air pressure.

Air pressure

Air pressure is caused by gravity. The Earth's gravity pulls all the air around the Earth down towards the ground. Although air is light, there is a lot of it, so it adds up to a fair amount of stuff bearing down on your head, and pushing in at the sides.

But you don't feel squashed, do you? This is because it is all around you, and inside you, and the pressure is equal in all directions. So we don't usually notice air pressure.

The human body is built to withstand air pressure. High in the sky, the air pressure is lower, and this can make people feel ill. Plane cabins are pressurized to keep the air pressure almost the same as at ground level.

Above: Air pressure pushes down on your hand, but it also pushes in from the sides and up from below. So you don't feel the air as a heavy weight.

The Earth's atmosphere

The atmosphere is the layer of air around our planet. It contains five thousand trillion (5 000 000 000 000 000) tonnes of air.

This diagram is not to scale. In real life, the atmosphere is a thin layer around the Earth.

Plane cabins lose a little pressure when they fly high. Can you think what the result is?
See page 30.

Above: The Breitling Orbiter is a hi-tech balloon designed for flying around the world high in the atmosphere. It has a pressurized chamber so that its pilots can have normal air pressure around them.

Water pressure

Divers deep beneath the sea experience the powerful force of water pressure. It is much greater than normal air pressure, as water is much heavier than air. Divers have to adjust slowly to the increase in pressure as they descend, and adjust again to the decrease in pressure as they rise back to the surface.

Sticky paper

Place a ruler on a flat table, with one end sticking out over the edge by about 2 cm. Lay a sheet of newspaper on top of the ruler and smooth it flat. When you tap the end of the ruler hard, you would expect it to flip upwards. But the paper seems to glue the ruler down to the table.

Can you think why the paper holds the ruler down so well? See page 30 for the answer.

Using air pressure

Aristotle (see page 4) mistakenly thought that flames reach upwards because they want to get back to the Sun. In fact, this happens because of air pressure.

As a flame burns it heats the air, making it lighter. The surrounding, heavier air is being pulled down harder by gravity and thrusts the flame upwards out of the way.

Hot-air balloons work because the air in them is hotter and lighter than the surrounding air. They are pushed up by the heavier air that surrounds them and sinks down beneath them.

Uses of air pressure

A suction pad (like the ones you get on bathmats) doesn't suck! In fact, it's the other way round. Because there is no air beneath the suction pad, air pressure keeps the pad pressed against the surface. The pad doesn't suck – the atmosphere pushes!

Right: The shape of a plane's wing makes air pressure push more on its underside than on its upper side. This shape is called an aerofoil.

Suckers and flames

When suction pads are not in use, air pressure pushes against them from all sides. When you press them onto something, you squeeze the air out from underneath them. No air can get in to push at them from the inside, so the air pressure outside holds them firmly against the surface.

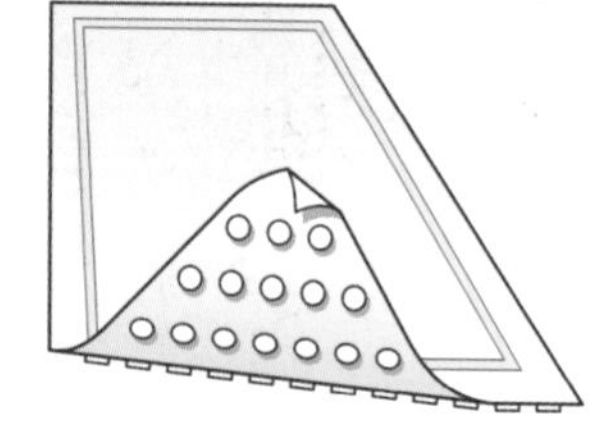

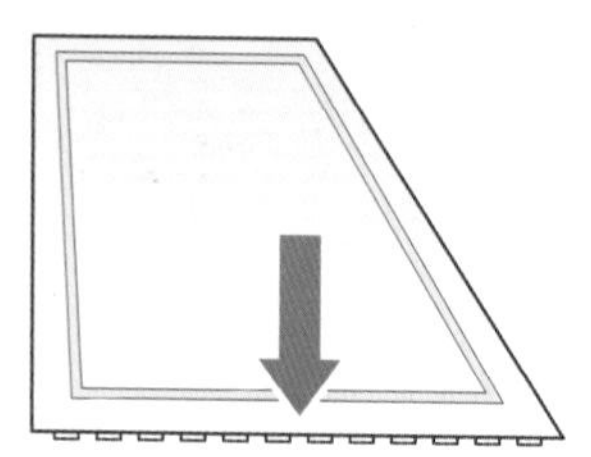

On Earth, air pressure makes flames point upwards. In a spacecraft, a candle would burn with a spherical flame, as there is no gravity to pull the air down around it and push it upwards.

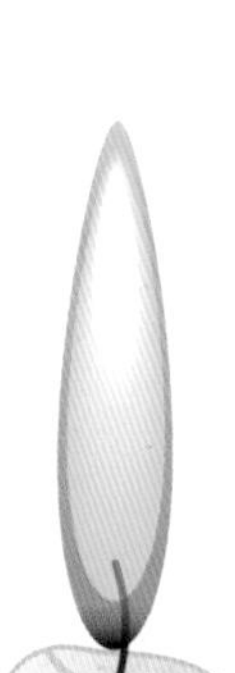

Plane and simple

Hold a piece of paper up flat in front of your mouth, and blow hard over it. Your blowing reduces the air pressure above the paper, and the greater air pressure below pushes it upwards.

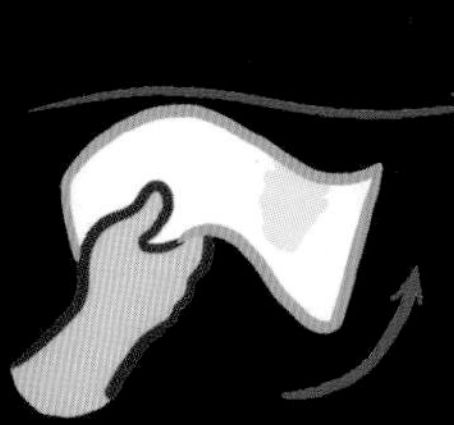

Here's another simple experiment that works the same way. Take two sheets of paper, and blow in between them. Instead of being pushed apart, they will be pushed together by air pressure.

Air pressure and flight

The cleverest use of air pressure is the aeroplane. When air moves fast, its pressure drops. In the first experiment above, because there is less air above the paper as you blow, the air pressure underneath pushes the paper upwards. Aeroplane wings are designed to make air flow over them faster than under them, reducing the pressure above the wing, so the force of the air below pushes the plane up in the air.

Birds' wings use the effects of air pressure to help them fly. It took humans a lot longer to work this out!

Above: A wind tunnel is a huge tunnel with a giant fan at one end to create a very strong wind. Wind tunnels are used to test the way air flows over plane wings.

Electrostatic force

Electrostatic force – also known as static electricity – is another invisible force. It is the force that makes a plastic comb pull bits of paper towards it after you've combed your hair.

Particle glue

Electrostatic force is what binds the **particles** of the universe together. For instance, the chemical carbon monoxide is made from **atoms** (tiny particles) of carbon and oxygen 'glued' together by electrostatic attraction to form a carbon monoxide **molecule**. The same is true for every molecule in the Universe.

Making molecules

This diagram shows how a water molecule is made from three atoms joined together.

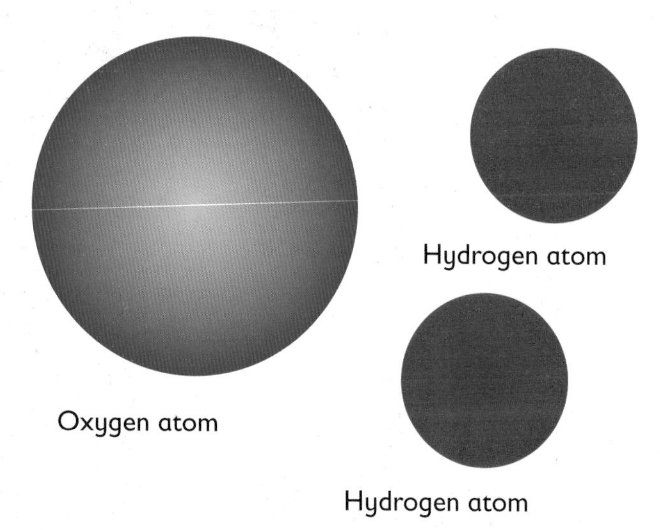

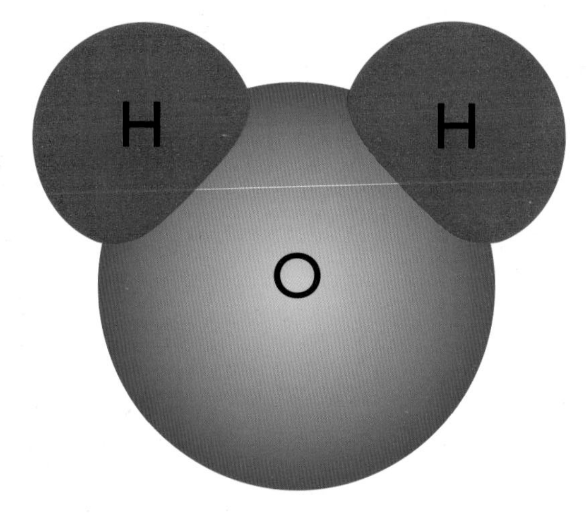

Electrostatic force holds them together.

Discovering electricity

The ancient Greeks noticed electrostatic force. They found that if you rubbed amber on cat fur, the amber would pull small objects towards it. But they did not know how it worked. It wasn't until 250 years ago that scientists started to study electrostatic force seriously.

Static electricity gets its name because it stays in one place. But when **Alessandro Volta** made the first battery in around 1800, it became possible to send this force down a wire as a flow of energy, which we call electricity.

Right: Lightning is a kind of static electricity.

Static experiments

Take a plastic comb and comb your hair with it. Then hold the comb near some small pieces of torn-up tissue paper. It should pick them up.

For another experiment, rub a blown-up balloon on a woolly jumper, then hold the same side of the balloon to a wall. Electrostatic force should make it stay there.

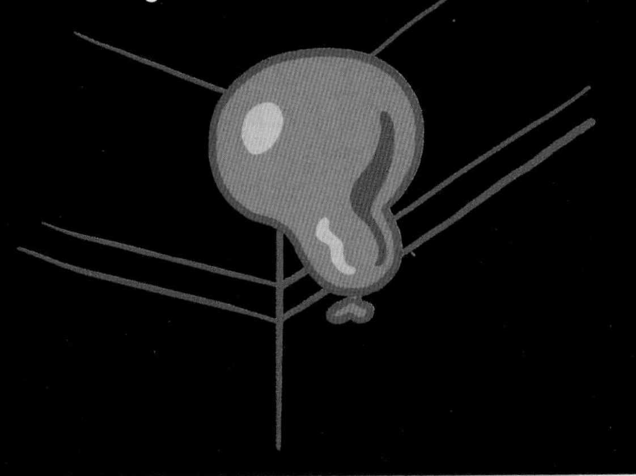

Magnetism

People have known about magnetism for thousands of years. The first magnets known to humans were lodestones. These are natural magnets that are sometimes dug up from the ground. People noticed that lodestones attracted iron objects towards them. And if you stroked an iron needle against a lodestone, the magnetic force was rubbed into the needle and it became magnetic too.

Magnets and electricity

In the 1820s scientists found that you could make a magnet simply by wrapping a wire around an iron bar and sending electricity through it. Could it be that the two forces, electricity and magnetism, were somehow related? Finally it was realized that electricity and magnetism were two faces of the same force – a force called electromagnetism.

Above: A compass contains a magnetized needle that can swivel freely. The needle tends to point north, because the Earth has is own magnetic force, which attracts one end of the needle. This makes magnets very useful for finding your way.

Changing north

Ask a grown-up to run a wire from a 12v battery across the face of a compass, directly on top of the needle. When the power is switched on, the needle turns aside slightly. This small movement shows that an electric current can make a magnet move. This discovery led to the electric motor (see opposite).

Left: A lodestone, or natural magnet, attracting a variety of metal objects.

Magnetic Earth

Magnets point north because the Earth is itself a vast magnet.

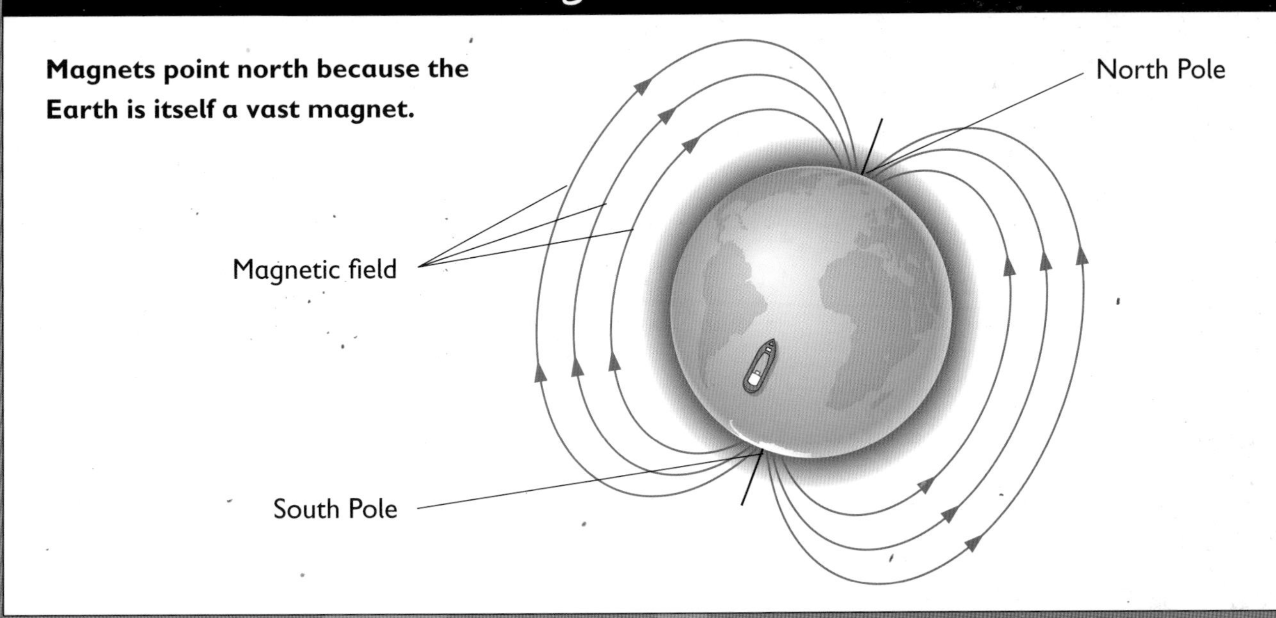

Making motors

In 1821 Michael Faraday gave a demonstration at the Royal Institution in which he combined a magnet and electricity. When an electric current ran through the wire, the wire rotated around a fixed magnet. This demonstration changed the world! It allowed people to build electric motors, which turn a flow of electricity into a rotating movement. Although it seemed like a small demonstration at the time, Faraday could see that it was of immense importance. Within 20 years electric motors were being used everywhere, to do the jobs that had previously been done by hot, messy, noisy steam engines.

Do you have a magnet in the house? See page 30.

Above: Michael Faraday with some of his scientific equipment.

Useful electromagnetism

In the modern world, we use electromagnetic force to help us with thousands of everyday jobs and industrial activities. There are two main inventions that rely on electromagnetism – the motor and the generator.

Motors

An electric motor uses a flow of electricity to make a magnet rotate. The rotating magnet is used to drive a shaft or wheel. Today every home has dozens of these machines. For example, there is a motor in a DVD player that makes the DVD spin around as it is played.

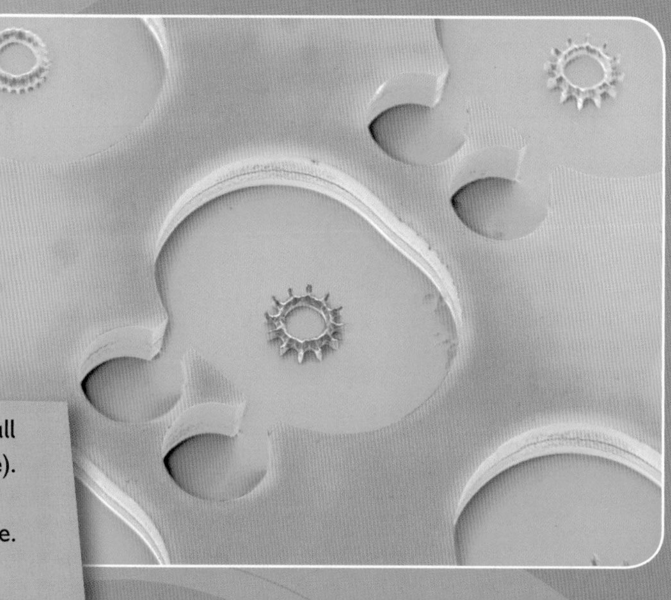

Right: Scientists have found ways to make incredibly small motors, or nano-engines (shown here under a microscope). In fact, the smallest one made so far is so tiny that 300 of them could fit onto the full stop at the end of this sentence.

Inside a hairdryer

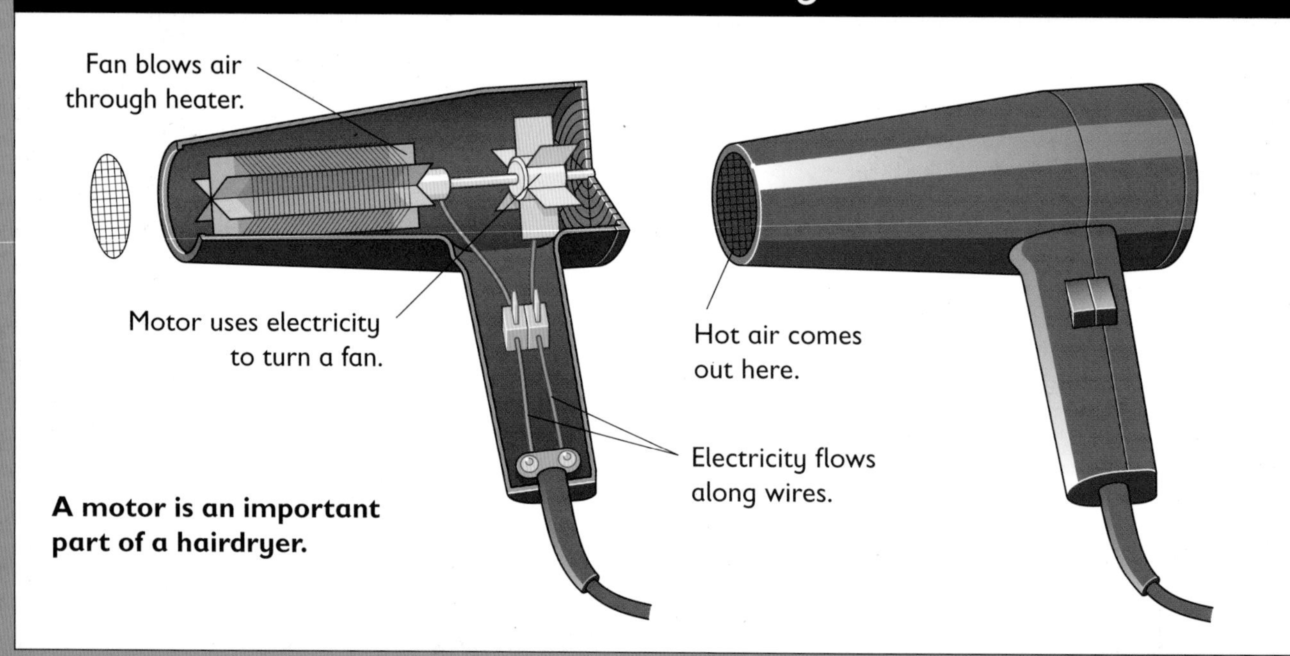

A motor is an important part of a hairdryer.

Make a nail magnetic

You will need magnet coiling wire from an electrical shop. Wrap it neatly round a nail as many times as you can. When you attach the ends of the wire to a 6v battery, the nail will become magnetic. (You can test it by holding a metal paperclip near it.) The more coils of wire there are, the stronger the magnetic force will be.

Generators

The science that makes a motor work can also be applied in reverse. A motor uses electricity to make a magnet rotate. But if you make a magnet rotate near an electric wire, it makes electricity flow in the wire. This is a generator. It means we can use rotating movement to produce, or generate, electricity. This is where most of our electricity supply comes from. In a power station, giant magnets spin rapidly to create huge amounts of electricity, which is carried along cables to wherever it is needed.

Right: Pylons hold up cables that carry electricity from power stations to homes and factories.

Here's a list of electrical devices. How many of them do you think have electric motors inside? See page 31.

Electric drill · Hairdryer · Electric fan · Food mixer · Fridge

Dimmer switch · Television · TV remote control · Computer

Electric doorbell · Vacuum cleaner · Washing machine · Telephone

?

Simple machines

Humans have always experimented with forces and used them to do things. To help with this, we have invented machines. A machine is an object that helps us use forces to do a particular job. Machines aren't always complicated or full of cogs and other bits. Even a simple stick can be a machine, if it's used in the right way.

Levers

One of the earliest machines was the lever – a stick balanced on a point, or pivot, so that it can move up and down like a seesaw. A large, weak movement at the long end of the lever produces a small, strong force at the short end. This is useful for lifting blocks to build city walls.

Alternatively, a powerful force at the short end of a lever can create a rapid movement at the other. This idea was used to make catapults, handy for knocking those city walls down again.

Right: A trebuchet was a giant catapult that worked using a lever system.

Everyday levers

All these everyday items use levers to make life easier.

Pliers

Bottle opener

Nutcracker

Scissors

Inclined plane

Another simple machine is the inclined plane – or, put more simply, a slope. It's very hard to lift a heavy weight up a big step. But if you put a slope there instead, you can push the weight up the slope much more easily. Why? Because you are moving the weight over a greater distance. You put in just as much effort in the long run, but you are not doing the work all at once. We still use inclined planes all the time. For example, a delivery driver might use one to roll heavy barrels in and out of a van.

Left: Rolling a ball up a slope is easier than lifting it straight up.

Screw slope

A screw is a machine that uses an inclined plane. The thread of the screw is like a slope running around and around it. As you turn a screw around and around to screw it into a piece of wood, you are making it travel much further than if you just tried to push it in. The effort is spread out, making it easier to do.

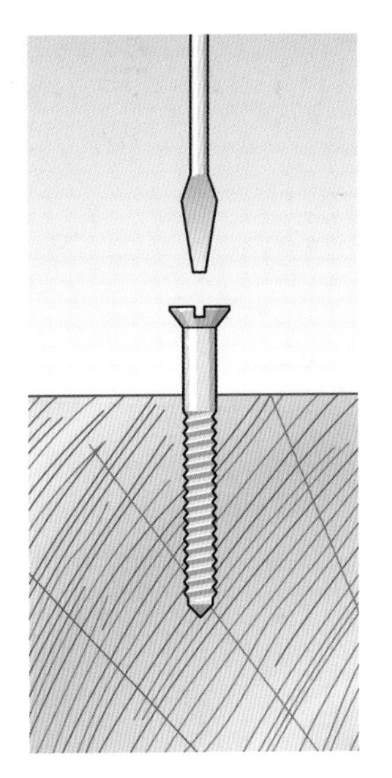

A ruler lever

Can you lift a chair with a ruler? Probably not. But try putting an eraser under the ruler near one end, and using it as a lever. When you press down on the long end, the short end will push up with great force.

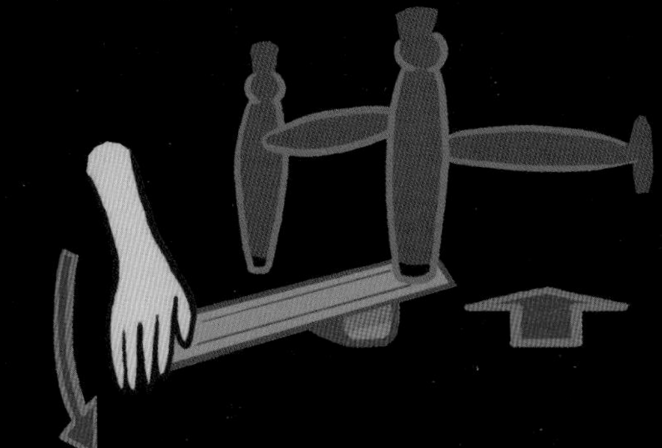

Now try using your ruler lever, as shown below, to flick buttons or paper balls.

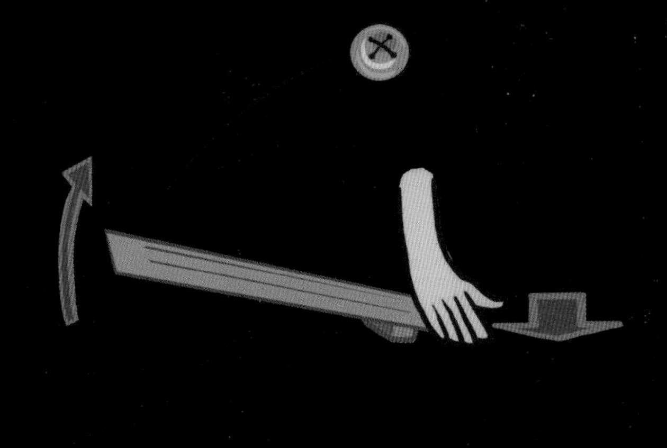

More machines

Many machines let you send a force from one place to another. Or they make a force change direction. This is called transmitting a force.

Everyday examples

Have you ever seen someone opening a very high window with a long pole? The pole carries, or transmits, the force you apply to the bottom of the pole to the window at the top. When you open an umbrella, you push upwards, and a set of levers pushes all the ribs outwards. The upwards force is changed into a sideways force. Have a look at an umbrella to see how it works.

Church bells are an example of one kind of force being changed into another. The bellringer pulls on a rope to ring the bell. The top of the rope pulls on a wheel that makes the bell rotate in a circle, so that its clapper moves and it rings. And when you open a door, a turning motion of the handle is changed into the straight movement of the catch. The diagram below shows how.

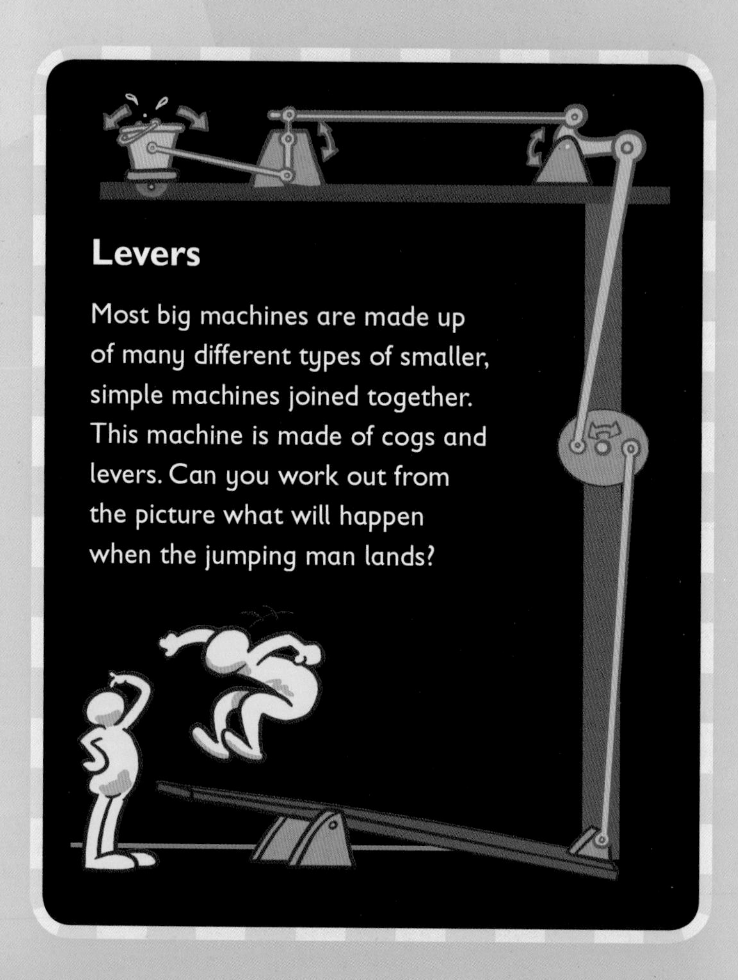

Levers

Most big machines are made up of many different types of smaller, simple machines joined together. This machine is made of cogs and levers. Can you work out from the picture what will happen when the jumping man lands?

Inside a door handle

A door handle turns rotating movement into sideways movement, using a part called a ratchet.

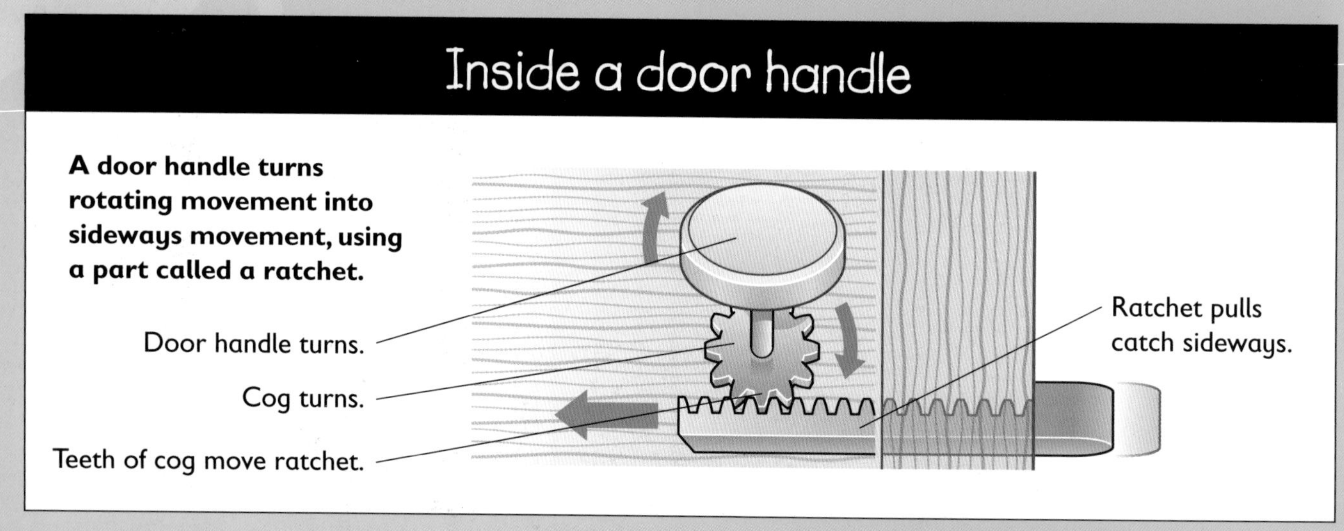

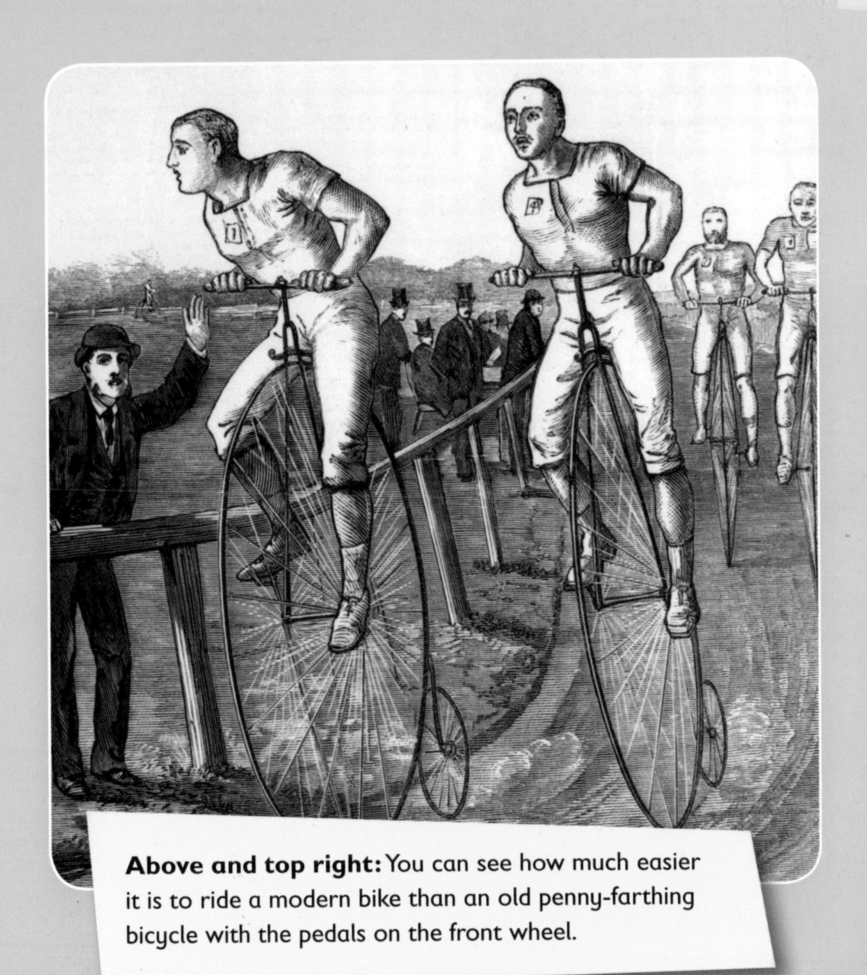

Above and top right: You can see how much easier it is to ride a modern bike than an old penny-farthing bicycle with the pedals on the front wheel.

On your bike!

The bicycle is a great example of transmitting forces. Some early bicycles had the pedals on the front wheel. You turned the pedals with your feet, and the wheel turned too. The problem was, the handlebars were also attached to the front wheel, for steering. They turned the wheel from side to side, making pedalling very difficult.

In later versions of the bicycle, the pedals were attached to the frame, below the seat. Then the pedal-power was transmitted to the back wheel, using a chain and a cog wheel. Bicycles are still made this way today.

Right: How a church bell works.

The engine

An engine is a machine that changes fuel into movement. Humans have used engines for over 200 years to power various types of transport, such as ships, trains and cars. The first engines were steam engines. Later, petrol engines, like those used in modern cars, were invented.

The steam engine

Early steam engines were built around a tube, or cylinder, with a block called a **piston** moving up and down inside it. The engine burned fuel which heated water, making steam. The steam filled the cylinder, pushing the piston up. When the steam cooled, it turned back into water. This reduced the pressure inside the cylinder, and air pressure pushed the piston down. More steam entered the cylinder, and the piston moved up again. The moving piston could be used to make a machine work.

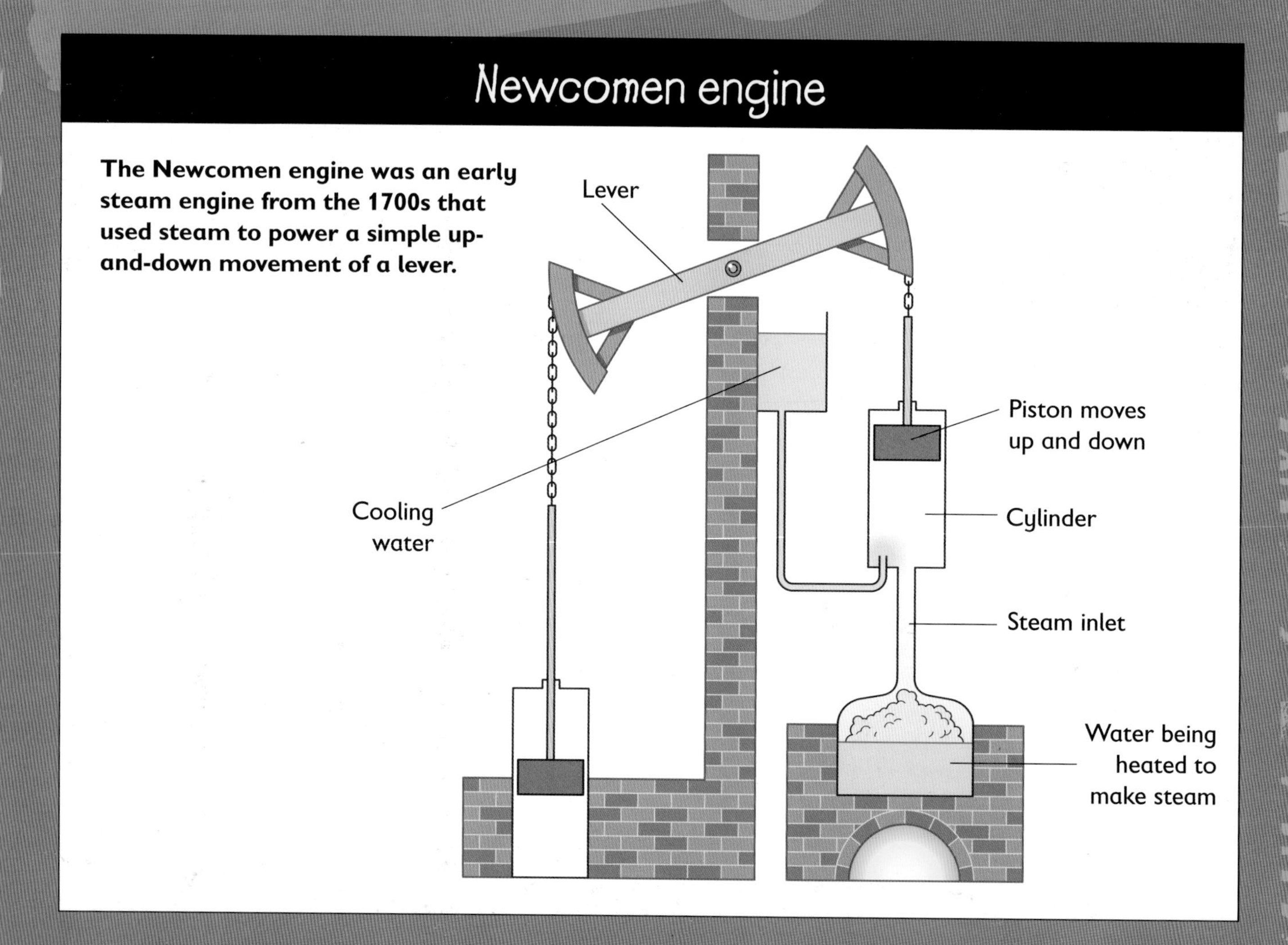

Right: Most modern vehicles, such as this snowmobile, are driven by a petrol engine.

The petrol engine

Petrol engines also have cylinders and pistons. They work by making a small amount of petrol explode inside the cylinder. This pushes the piston outwards. Between explosions, the piston moves back in. In a car, the back-and-forth motion of the pistons makes the **axle** and wheels rotate.

Below: When steam trains were first invented in the 1820s, many people were frightened when they travelled on them.

Old forces for a new age

At the start of the 21st century, the world uses vast amounts of natural fuels – such as oil, gas and coal. They power our cars and planes and provide us with electricity to run our hairdryers, food mixers, fans and millions of other appliances. We may have to do things differently soon, when these fuels run out.

Back to the future

To find new sources of energy, we may have to turn to natural forces. People long ago used many of these to provide energy, and now we may need them again. For example, the flow of a river, pulled downhill by gravity, can be used to turn a waterwheel. In the past these provided power for factories and mills. Today, the force of flowing water is turned into electricity in hydroelectric power plants. Windmills once used the wind to power grain-grinding equipment. Today, modern windmills, or wind turbines, use the wind to generate electricity.

Above: Machines use forces to make our lives easier, but many of them need an energy supply.

Left: Wind farms such as these are increasing in number as a way of generating electricity without using up fuel.

Above: A hydroelectric dam channels the water in a river so that it can be used to run a power station.

More sources of force

The Moon's gravity pulls on the Earth's seas and oceans, causing the tides. The force of the tides can be harnessed by **tidal generators**.

There is another force we can use, too – the force that holds atoms together. Nuclear power releases energy by splitting atoms open.

As the Earth is a giant magnet, a giant wire, circling the planet in space, could generate electricity. This may be the energy source of the future.

We have only had electricity for 200 years and steam engines for 300 years, yet we feel totally dependent on the forces they can provide. How would your life change if there were no engines? See page 31.

Questions and answers

Do you know of any other strange things ancient peoples used to believe about the Earth? (page 5)

They used to think that the Earth was flat, and that the Sun, Moon and planets revolved around it.

If it wasn't for friction, a game of snooker would be impossible. The balls would roll around the table for ever. How do you think a friction-free snooker game would end? (page 8)

The balls would keep moving around the table until they all fell into the pockets.

Even though you are tiny compared to the Earth, you have your own gravitational field. While the Earth is pulling you down, you are very slightly pulling the Earth up. Can you think of something that has enough gravity to have a bigger effect on the Earth? (page 10)

The Moon is very large, though still much smaller than the Earth. Its gravity has an effect on the Earth by pulling at the oceans and creating tides.

Plane cabins lose a little pressure when they fly high. Can you think what the result is? (page 12)

Your ears pop, because the pressure inside them becomes greater than the pressure outside, and your eardrums are pushed outwards.

Can you think why the paper holds the ruler down so well? (page 13)

Because when you hit the ruler, air cannot get under the paper quickly. Where the paper is flat against the table, there is no air pressure pushing it up, and the weight of the atmosphere holds it down.

Do you have a magnet in the house? (page 19)

TVs, drills and some screwdrivers contain magnets. In fact, anything with a motor in it contains a magnet. There might be some on your fridge door, too.

Here's a list of other electrical devices. How many of them do you think have electric motors inside? (page 21)

All of these contain one or more motors: electric drill, hairdryer, electric fan, food mixer, fridge, computer, vacuum cleaner, washing machine.

We have only had electricity for 200 years and steam engines for 300 years, yet we feel totally dependent on the forces they can provide. How would your life change if there were no engines? (page 29)

To help answer this, think about how people used to live in the 1700s. You can look at old paintings to give you an idea.

Glossary

Aristotle (384–322 BCE) – Perhaps the first scientist, Aristotle pointed out that natural events were not caused by gods. Although his ideas about what did cause events were wrong, he is the one who started the search for scientific truth.

Atmosphere – The layer of gases surrounding the Earth.

Atoms – Tiny particles that all matter is made up of.

Axle – A shaft that the wheels in a vehicle are connected to, allowing them to turn.

Eclipse – The blocking of the Sun's or Moon's light when the Moon or Earth is in the way.

Energy – The power something has to work.

Force – A push or pull.

Gravitational field – The area surrounding an object in which gravity (see below) works.

Gravity – A force that makes objects pull towards each other.

Mass – The mass of an object is the quantity of matter (or stuff) in it.

Molecule – A particle made up of atoms joined together. For instance, a molecule of water is made of atoms of hydrogen and oxygen.

Newton, Isaac (1643–1727) – One of the greatest scientists in the world. His lifetime marked the final dismissal of Aristotle's 2000-year-old ideas of science and the start of the modern era. He made extraordinary discoveries about light and gravity, as well as making clear how forces work.

Particle – A tiny part or unit, such as an atom or a molecule.

Piston – Part of an engine that is moved up and down inside a cylinder by the pressure of steam or burning gases. The piston is connected to a shaft that drives movement of a machine, such as a car.

Re-entry – The return of a spacecraft from space back into the Earth's atmosphere.

Tidal generators – Machines that use the power of the tides to generate electricity. As the tide rises water pours into a large container. At high tide the gates are shut and the water is kept there until the tide has gone out. At low tide the water is released through a turbine, and makes electricity as it pours out.

Volta, Alessandro (1745–1827) – An Italian count who made many discoveries. His most important invention was the 'voltaic pile' or battery – the first time electricity could be made to flow in a steady stream.

Index